BEGINNL
GARLIC FOR PROFIT

Elevate Your Garlic Growing Game with In-Depth Insights into the Life Cycle, Pest Management, and Post-Harvest Handling for Flavorful and Healthy Garlic Bulbs

Lucille Sarron

Table Of Contents

CHAPTER 1 6
  Garlic 6
    An Overview Of Growing Garlic: 6
    Demand And Importance In The Market: 8
    Possibility Of Financial Gain: 9
  CHAPTER 2 12
  Recognizing Different Types Of Garlic 12
    Popular Types Of Garlic 12
    Selecting The Appropriate Type For Your Area
    14
    Properties And Taste Profiles 15
  CHAPTER 3 18
  Getting Ready For Soil And Planting 18
    Optimal Garlic Soil Conditions: 18
    Testing And Amendments For Soils: 19
    Planting Methods: 20
  CHAPTER 4 22
  Environmental And Climate Factors 22
    Climate And Temperature Requirements: 22
    Needs For Water And Sunlight: 23

Defense Against Diseases And Insects: 24
CHAPTER 5 28
Upkeep And Scenario 28
Embryology Procedures: 28
Strategies For Weed Control: 29
Handling Scapes Of Garlic: 31
CHAPTER 6 34
Garlic Harvesting 34
1. Indices Of Adulthood: 34
2. Methods Of Harvesting: 35
3. How To Cure And Store Garlic: 36
CHAPTER 7 38
Management of Diseases and Pests 38
Natural Control: 39
Chemical Regulation: 40
Steps To Prevent Disease: 41
CHAPTER 8 44
How To Promote And Sell Your Garlic......44
1. Choosing The Market You Want To Enter: 44
2. Packaging And Branding Techniques: 45
3. Platforms & Channels For Marketing: 46

- Summary 48
  - Summary Of The Main Ideas: 49
  - Motivation For Upcoming Achievements: 50
  - Ongoing Education And Development: 51
- THE END 53

CHAPTER 1

Garlic

For ages, people have grown garlic (Allium sativum), a multipurpose and extensively used plant with both culinary and medical uses. Due to its unique flavor and many health advantages, garlic is becoming a common element in many different cuisines worldwide. This tutorial examines the complexities of garlic farming, focusing on the crop's significance, market demand, and financial possibilities.

An Overview Of Growing Garlic:

1. Garlic Varieties:

- Softneck garlic: This type is good for braiding because of its extended shelf life and mild flavor.

- Hardneck garlic: Preferred by foodies, hardneck garlic varietals are known for their strong flavor and bigger cloves.

2. Conditions of the Soil and Climate:

- Fertile, well-drained soil with a pH of 6.0 to 7.5 is ideal for garlic growth.

Garlic grows well in a cold climate with plenty of sunshine.

3. Sowing and Spreading:

- Cloves, which are the separate parts of the garlic bulb, are how garlic is spread.

- Plant cloves in the autumn so they can get established before the winter.

4. Cultivation Methodologies:

- To guarantee optimal bulb growth, planting depth and spacing must be done correctly.

- Mulching inhibits the growth of weeds and preserves soil moisture.

- For the best development, regular fertilizer and watering are essential.

5. Management of Pests and Diseases:

Aphids and nematodes are frequent pests, and diseases like rust and white rot can harm garlic plants.

Crop rotation and the use of organic pesticides can also assist to lessen these problems.

Demand And Importance In The Market:

1. Taste-Related Significance:

- An essential component of many different cuisines, garlic improves the flavor of foods including marinades, soups, and sauces.

2. Therapeutic Qualities:

- Garlic has been linked to several health advantages, such as the ability to decrease blood pressure, and cholesterol, and strengthen the immune system.

3. Usages in Culture and Tradition:

• Garlic is used in cooking and medicine, but it also has cultural and traditional value in many countries.

Possibility Of Financial Gain:

1. Trends in the Market:

Increasing demand for locally sourced and organic goods offers garlic producers a chance to enter specialized markets.

2. Items with Additional Value:

• Product diversification may boost revenue. Examples of this include making pickled garlic, infused oils, and powdered garlic.

3. Sales directly to consumers:

• Direct customer ties may be built through internet platforms, farmers' markets, and community-supported agriculture (CSA). This can increase profitability.

4. Expense Control:

- Increasing profit margins may be achieved by using cost-effective technology, maximizing resource utilization, and implementing efficient agricultural methods.

5. Branding and Quality Assurance:

Growers may command premium pricing in the market by building a strong brand and regularly producing high-quality garlic.

In conclusion, cultivating garlic gives the joy of producing a product that is both adaptable and in demand, as well as the possibility of financial gain through value addition, marketing, and strategic planning. Garlic farming needs a mix of expertise, knowledge, and a deep awareness of market dynamics to be successful, just like any other agricultural endeavor.

CHAPTER 2

Recognizing Different Types Of Garlic

Allium sativum, or garlic, is a popular and adaptable culinary item prized for both its unique taste and plethora of health advantages. Knowing the differences between the types of garlic is essential for producing a good crop. Garlic comes in two primary varieties: hardneck and softneck, each with several subvarieties.

Popular Types Of Garlic

1. Hardneck Types:

• Rocambole garlic: Chefs praise Rocambole garlic for its easy-to-peel cloves and rich, deep flavor. It generates bulbils on a tall blooming stalk known as a scape.

• Porcelain: The flavor and size of porcelain garlic cloves are well-known. It's a popular option among producers because of its lengthy storage life and gorgeous white skin.

• Purple Stripe: The bulb wrappers of this cultivar have eye-catching purple stripes. Purple Stripe garlic usually contains fewer cloves per bulb and a strong, spicy taste.

2. Types of Softnecks:

• Artichoke: Artichoke garlic has a mellow, moderate flavor and is a softneck kind. It is well-known for having superior storing properties and often contains more cloves per bulb.

• Silverskin: This softneck garlic cultivar has tiny cloves that are closely spaced. Its supple stems make it a popular choice for braiding, and it tastes robust and spicy.

Selecting The Appropriate Type For Your Area

Choosing the right garlic type for your area is essential to a good yield. Take into account the following elements:

1. climatic: The temperature and climatic requirements of various garlic cultivars vary. For instance, hardneck types tend to do better in colder climes, whilst softneck variants are more suited to warmer temperatures.

2. Soil Conditions: Fertile, well-draining soil is ideal for garlic. Knowing your soil type is important since different cultivars may have different preferences for certain types of soil.

3. Day Length: Certain types of garlic work better throughout particular day lengths. For example, certain types could need a longer duration of cold weather in order to initiate bulb production.

4. Local Adaptation: Certain local cultivars may perform better than others since they have spent time adapting to your particular location. To find out which kinds are effective in your region, speak with knowledgeable growers or the local agricultural extension agencies.

Properties And Taste Profiles

Garlic's flavor varies greatly throughout kinds, which might affect how your food turns out. Think about the following qualities:

1. Gentle vs. Robust: Softneck cultivars often possess a more delicate flavor, rendering them appropriate for ingestion uncooked or in recipes that call for a delicate hint of garlic flavor. Hardneck cultivars frequently offer more robust, nuanced flavors.

2. Sweet vs. Spicy: While some garlic kinds taste sweeter, others have hotter undertones. Knowing

these taste qualities will assist you in selecting the appropriate garlic for a certain dish.

3. Storage Properties: Some types store information better than others. It's crucial to choose a garlic type with exceptional keeping properties if you intend to preserve it for a long time.

To sum up, growing garlic successfully requires choosing the appropriate cultivars based on your area, environment, and cooking style. By trying out several varieties, you may learn about their subtleties in flavor and adaptability, which will improve both your culinary creations and gardening experience.

CHAPTER 3

Getting Ready For Soil And Planting

Successful garlic cultivation necessitates meticulous consideration of planting techniques and soil preparation. Here is a detailed explanation of all the important ideas for preparing the soil and planting garlic:

Optimal Garlic Soil Conditions:

1. Well-Drained Soil: To avoid wet circumstances that might cause rot, garlic needs well-drained soil. Loamy or sandy loam soils work best because they retain moisture and allow for good drainage.

2. pH Levels: Soils that range from 6.0 to 7.5 are ideal for garlic growth and are mildly acidic to neutral. You may use lime to raise pH or sulfur to reduce it after doing a soil test to find out what your soil's pH is.

3. Rich in Organic Matter Fertile Soil: Fertile soil is beneficial for garlic growth. Before planting, add compost or well-rotted manure to the soil to enhance soil structure and supply vital nutrients.

Testing And Amendments For Soils:

1. Soil Testing: To ascertain the precise amendments required for the best possible development of garlic, do a soil test to evaluate pH and nutrient levels.

2. Garlic needs nitrogen to produce leaves, and it also needs potassium and phosphorus to form bulbs. Add organic additions like bone meal and potassium sulfate, or a balanced fertilizer, to the soil in accordance with the findings of the soil test.

3. Organic Matter: Adding organic matter will improve the structure and fertility of the soil. To enhance the texture and nutrient content of the soil, add cover crops, well-rotted manure, or compost.

4. Micronutrients: Identify any inadequacies in these areas and take appropriate action. Zinc and boron are two common deficits that can be addressed with the application of appropriate micronutrient supplements.

Planting Methods:

1. Choosing Bulbs: When planting garlic bulbs, make sure they are healthy and free of diseases. Split the bulbs into cloves, making sure that each clove is healthy and full.

2. Garlic cloves should be planted with their pointy ends facing up. Planting at a depth of around 2 inches (5 cm) is good. Planting bulbs too deeply or too shallowly might impact their growth.

3. Planting: Arrange cloves in rows 12–18 inches (30–45 cm) apart, with individual cloves spaced about 6–8 inches (15–20 cm) apart. Appropriate spacing promotes proper air circulation, which lowers the chance of illness.

4. Mulching: After planting, cover the soil with an organic mulch, such as chopped leaves or straw. Mulching aids with weed suppression, moisture conservation, and winter garlic insulation.

5. Watering: Ensure that the soil is continuously wet, particularly in the early phases of development. Watering is essential to the growth of strong roots and bulbs.

6. Seasonal Timing: Sow garlic a few weeks ahead of the first heavy frost in the fall. This facilitates the garlic's establishment of roots prior to winter and encourages vigorous growth in springtime.

In conclusion, thorough planting and soil preparation are the first steps in a good garlic crop. You may guarantee a plentiful crop of nutritious and delicious garlic bulbs by applying appropriate planting techniques, testing the soil, and establishing the right soil conditions.

CHAPTER 4

Environmental And Climate Factors

Allium sativum, or garlic, is a resilient crop that grows well in a range of temperatures. However, it's important to pay attention to certain climatic and environmental factors in order to secure a successful harvest. Here's a detailed explanation of temperature, sunshine requirements, watering schedules, and pest and disease prevention for garlic plants:

Climate And Temperature Requirements:

1. Range of Temperature:

As a crop that grows best in chilly weather, garlic prefers temperatures between 55°F and 77°F (13°C and 25°C).

- Wintertime lows are crucial for bulb development because exposure to the cold initiates bulb growth.

2. Need for Chilling:

- For healthy bulb growth, garlic needs a period of low temperatures (around 32°F or 0°C) for a specific number of weeks. The amount of chilling required varies depending on the type of garlic.

3. Flexibility:

- A broad variety of temperatures, from cold temperate to subtropical, may support garlic growth. It does, however, function best in areas with diverse seasons.

Needs For Water And Sunlight:

1. Sufficient Sunlight:

- Garlic needs at least six to eight hours of direct sunshine each day in order to thrive.

- Sufficient sunshine encourages photosynthesis, which results in strong growth and bulb development.

2. Moisture Content of Soil:

- Soil that drains well is essential to avoiding wet circumstances that might cause decay.

- Sufficient moisture is necessary for healthy growth, particularly when bulbs are forming. On the other hand, too much moisture can cause garlic to rot.

3. Watering Schedule:

- In general, garlic needs regular irrigation, particularly in dry spells.

- Creating a watering regimen that keeps the soil continuously damp but not soggy is essential.

Defense Against Diseases And Insects:

1. Typical Pests:

• Pests including aphids, thrips, nematodes, and onion maggots can harm garlic.

• Using natural predators or organic insecticides in conjunction with routine monitoring can help control pest infestations.

2. Prevention of Diseases:

• Diseases including rust, downy mildew, and white rot can affect garlic.

• Some diseases may be avoided by rotating crops, ensuring that the soil drains properly, and avoiding overhead irrigation.

3. Planting companion plants:

• Garlic may be planted with companion plants, such as marigolds, to help ward against pests and enhance general plant health.

4. Fungal Problems:

- It's critical to maintain adequate air circulation around the plants and steer clear of overcrowding in order to prevent fungal diseases like white rot.

To sum up, cultivating garlic effectively entails knowing and controlling the temperature needs, giving the crop enough sunshine and water, and putting good pest and disease control measures in place. Garlic will grow abundantly if these factors are tailored to your unique climate and environmental circumstances.

CHAPTER 5

Upkeep And Scenario

Effective garlic growing necessitates close attention to a number of maintenance and care requirements. We'll go over important topics including fertilization techniques, weed management tactics, and handling garlic scapes in this conversation.

Embryology Procedures:

1. Getting the Soil Ready:

• Fertile, well-drained soil is ideal for garlic growth. To increase the soil's structure and nutrient content, amend the soil with organic matter, such as compost, before planting.

• Test the soil to ascertain the precise nutritional requirements. In general, garlic has a pH that is slightly acidic to neutral.

2. Selecting Appropriate Fertilizer:

• Choose a fertilizer that is well-balanced, having greater amounts of potassium, phosphorus, and mild nitrogen. A 10-10-10 or a comparable formulation is a common option.

• Organic substitutes, like as bone meal or well-rotted manure, can also supply vital nutrients.

3. When fertilization occurs:

• Before planting, apply fertilizer and fully incorporate it into the soil. By doing this, you can be sure the garlic cloves will have access to nourishment as they develop roots.

• When the garlic begins to develop aggressively in the early spring, apply a side dressing of fertilizer. Fertilize your bulbs early in the season to avoid postponing their maturation.

Strategies For Weed Control:

1. Mulching:

- Surround garlic plants with a layer of organic mulch (straw, hay, or crushed leaves). Mulching controls soil temperature retains soil moisture, and inhibits the development of weeds.

- Make sure the mulch is distributed uniformly to avoid nutritional competition.

2. Manual Weeding:

- Check the garlic bed frequently for weeds and pull them by hand. Since garlic is more susceptible to competition in its early development stages, this is especially crucial.

3. Cultivation Methodologies:

If you want to stop weeds from growing without harming garlic roots, use shallow cultivation. Because the roots of garlic are close to the soil's surface, exercise caution when planting.

4. Preventive actions:

- Take into account pre-planting weed control techniques including sterilizing equipment and utilizing planting materials free of weeds. This aids in lowering the soil's weed seed bank.

Handling Scapes Of Garlic:

1. Recognition:

The green, curling branches that grow from hardneck garlic cultivars are known as garlic scapes. To divert the plant's energy into bulb growth, these should be eliminated.

2. When to Remove Something:

- Late spring to early summer is when most scapes are ready to be removed. To keep them from turning woody, pull them off after they curl a single or two times.

3. Gathering Branches:

Approximately one inch above the uppermost leaf, cut the scapes using sharp scissors or pruning shears. Scapes that have been harvested are edible and have a little garlic flavor when cooked.

4. Effect on the Size of the Bulb:

• The removal of scapes is essential to maximize bulb size. When scapes are allowed to blossom, bulb growth is impeded, leading to smaller cloves.

In conclusion, a comprehensive approach to upkeep and care is necessary for good garlic growing. Garlic scape management, appropriate fertilization techniques, and weed control tactics are all tools producers may use to guarantee a plentiful crop of tasty, healthy garlic bulbs. A regular watch and prompt action plan can make the garlic growing endeavor a success overall.

CHAPTER 6

Garlic Harvesting

1. Indices Of Adulthood:

When the upper leaves of garlic remain green but the lower leaves begin to turn yellow or brown, it's usually time to harvest. When assessing maturity, keep an eye out for these particular indicators:

• Leaf Color: Keep an eye on how the leaves change in color. A mature plant will have between one-third and half of its leaves turn yellow or brown.

• Scapes: The garlic plant is directing its energy away from bulb development as scapes, or flower stalks, begin to form. Cut off scapes to return energy to the lightbulb.

• Examine Bulbs: Using a gentle digging motion, determine the size of each bulb and if separate cloves have developed. Well-formed but not excessively developed bulbs are what you want.

2. Methods Of Harvesting:

• Timing: Depending on the local growing circumstances and garlic type, harvesting times vary. Usually, it happens in the early summer or late spring.

• Digging: Gently loosen the dirt surrounding the bulbs using a shovel or garden fork. Take caution when digging so as not to harm the bulbs.

• Lift and Clean: Carefully shake off any extra dirt after lifting the garlic bulbs out of the ground. Removing the plants by their crowns might harm the bulbs, so avoid doing that.

• Curing: To improve flavor and storage capacity, let the picked garlic cure for a few weeks.

3. How To Cure And Store Garlic:

• Curing Conditions: The process of drying and solidifying garlic's outer layers in preparation for long-term preservation is known as curing. After harvesting, store the garlic away from direct sunlight in a warm, dry, and well-ventilated place.

• Bundle or Braid: To create an eye-catching presentation, tie garlic bulbs in bundles or braid the leaves together after they have cured for a week or two. Place them where there is good ventilation.

• Pruning: Following curing, prune roots and trim the back stem to approximately one inch above the bulb. This lessens the chance of mold growth and enhances the look of the garlic.

• Storage: Keep cured garlic in a well-ventilated, cold, and dark area. Garlic that has been properly preserved can be kept for several months.

Extra Advice:

• Prevent Over-Maturity: Garlic should be harvested before it is too mature since this can cause the cloves to split and shorten their shelf life.

• Test Bulbs: Slice a few bulbs open to ensure they are mature. The cloves must have distinct segments and firm skin.

• Frequent Monitoring: Keep an eye on the garlic patch as there may not be much of a harvesting window and timing is essential for best quality.

You may reap the rewards of your gardening efforts with tasty, long-lasting garlic bulbs if you follow these tips for harvesting, curing, and storing garlic.

CHAPTER 7

Management of Diseases and Pests

Although growing garlic is a pleasant effort, pests, and illnesses may affect it just like any other crop. Good management practices are essential to robust garlic production and abundant yield. We will discuss typical garlic pests in this setting, along with chemical and organic management strategies and disease-preventive tactics.

Typical Pests of Garlic:

1. Thrips: These little, thin insects cause discoloration and reduced development on garlic leaves as a result of their feeding.

2. Aphids: Aphids have the ability to infest garlic, sucking in leaf sap and spreading viruses.

3. Nematodes: Garlic roots can be harmed by soil-dwelling nematodes, which can hinder nutrient uptake and cause stunted plants.

4. Garlic bulbs are susceptible to the deadly fungus known as "white rot," which results in fluffy, white growth that eventually decays.

5. Onion Maggots: By feeding on garlic bulbs, onion fly larvae can harm garlic bulbs.

Natural Control:

1. Neem oil is an organic insecticide that works well against aphids and thrips. It throws off these bugs' eating and reproductive cycles.

2. Beneficial Insects: To naturally reduce aphids, introduce predatory insects like ladybugs and lacewings.

3. Companion Planting: You may keep pests away from garlic by planting it alongside companion crops like chives and marigolds.

4. Diatomaceous Earth: When diatomaceous earth is sprinkled over garlic plants, it forms a barrier that weakens bugs' exoskeletons.

5. Crop Rotation: Garlic may be planted in rotation with other crops to aid nematodes and other soil-borne pests disrupt their life cycle.

Chemical Regulation:

1. Insecticidal Soaps: These soaps work well against pests with soft bodies, such as aphids. They cause damage to insects' cell membranes.

2. Pyrethrin-Based Insecticides: Pyrethrin is a naturally occurring insecticide that works well against a range of pests. It is obtained from chrysanthemum flowers.

3. Systemic Fungicides: Applied to the soil or foliage, systemic fungicides offer long-lasting protection against diseases such as white rot.

4. Nematicides: To manage nematode infestations in the soil, apply chemicals such as fenamiphos.

Steps To Prevent Disease:

1. Good Sanitation Practices: To stop the spread of illnesses, remove and destroy contaminated plant material as soon as possible.

2. Plant Spacing: To encourage air circulation and lower the danger of fungal infections, make sure there is enough space between garlic plants.

3. Maintaining correct organic matter levels and a well-draining soil structure will help prevent the growth of soil-borne illnesses.

4. Choose garlic cultivars that are recognized for their ability to withstand common illnesses that are frequent in your area.

5. Crop monitoring: Check plants often for indications of pests or diseases, and respond appropriately when problems arise.

In conclusion, good pest and disease management in garlic farming may be achieved by a mix of preventative measures, organic controls, and the sparing application of chemical treatments. By putting these tactics into practice, you can make sure that your garlic crop is healthy for the long run and that the harvest is successful.

CHAPTER 8

How To Promote And Sell Your Garlic

1. Choosing The Market You Want To Enter:

Success in the garlic industry depends on knowing your target market. Take into account variables like psychographics, location, and demography. Determine who is most likely to purchase your garlic: specialist shops, eateries, grocery stores, or local residents. To determine the level of demand for garlic in your chosen region, conduct market research.

• Local Customers: Take into account things like their tastes, purchasing patterns, and the particular kind of garlic they are drawn to if your goal is to sell garlic to local customers directly.

• Restaurants and Culinary Businesses: If you are targeting these industries, you should know how much they need, what kinds of garlic they like, and if they would rather use processed or fresh garlic goods.

• Grocery Stores and Specialty Markets: Find out what packaging and quality requirements they have for products sold in these establishments. Recognize the competitors in these markets and how to differentiate your garlic.

2. Packaging And Branding Techniques:

You may differentiate your garlic from the competition by developing a distinctive brand identity and attractive packaging.

• Brand Identity: Create a distinctive brand identity that conveys the superiority and individuality of your garlic. Think of elements like a recognizable logo, an engaging brand narrative, and a recurring visual motif.

• Packaging: Make an investment in eye-catching and useful packaging. Make sure to specify the kind of garlic, where it came from, and any special features. To win over ecologically sensitive customers, think about using eco-friendly packaging solutions.

• Differentiation: Emphasize the unique qualities of your garlic. Use differentiators in your branding and packaging, such as a unique variety, organic certification, or a particular growing technique, to draw in buyers.

3. Platforms & Channels For Marketing:

Selecting the appropriate platforms and marketing channels is essential to successfully reaching your target audience.

• Farmers' Markets: Attend your community's farmers' markets to establish a direct relationship with customers. Provide samples, tell your tale, and establish a rapport with clients.

• Online channels: To reach a larger audience, make use of e-commerce channels. Make a user-friendly website where clients can shop from you, find out more about your business, and even sign up for recurring deliveries.

• Social networking: Use social networking sites to interact with clients, offer recipes, and exhibit your garlic. Visual platforms like Instagram and Pinterest may effectively highlight the attractiveness of your garlic.

• Partnerships: Take into account partnering with regional chefs, food bloggers, or influencers who might include your garlic in their posts. Word-of-mouth marketing and referrals from reliable sources may have a big influence on your sales.

• Community Engagement: Participate in collaborations, events, or sponsorships to give back to the community. Take part in culinary events or form partnerships with nearby companies to raise your profile.

In summary, developing a strong brand, knowing your target market, and selecting the best distribution methods are all essential to effectively marketing and selling garlic. To achieve long-term success, keep adjusting your strategy in response to feedback and market developments.

Summary

To sum up, cultivating garlic is a satisfying and rather simple process that may provide a plentiful and delicious harvest. We have discussed many facets of garlic growth throughout this examination of fundamental ideas, ranging from choosing the best garlic type to preparing the soil, planting, and preserving ideal growing conditions. As you begin or continue your garlic-growing adventure, it is important to review the main ideas that were covered.

Summary Of The Main Ideas:

1. Garlic types: The key to a successful harvest knows the many types of garlic. Each kind of garlic, whether softneck or hardneck, has distinct qualities, tastes, and cultivation requirements.

2. Soil Preparation: Garlic growth is greatly influenced by the quality of the soil. Make sure the soil is well-draining and rich in organic materials. Healthy garlic bulbs are the result of properly prepared soil prior to planting.

3. Planting Techniques: Be mindful of planting time, depth, and spacing. To ensure sturdy plants in the spring, plant cloves in the autumn and give them time to form roots before winter.

4. Maintenance Procedures: Weeding, mulching, and regular irrigation are all crucial maintenance

activities. Fertile garlic plants want constant moisture and an absence of weeds.

5. Harvesting and Curing: When it comes to harvesting garlic, timing is everything. When the lower leaves start to turn brown, gently remove the bulbs. The bulbs will reach their maximum taste and storage potential if they are cured properly and have adequate air circulation.

Motivation For Upcoming Achievements:

Taking on the task of producing garlic is more than just planting a crop; it's a chance for ongoing education and personal development. When you face obstacles, be they weather-related, pest-related, or unplanned setbacks, see them as learning opportunities that advance your knowledge as a garlic farmer. Accept the learning process and make the most of every opportunity to improve your abilities.

When you harvest your own garlic and see it grow from a little clove to a tasty bulb, you get an unmatched sense of accomplishment. Your dedication to using organic and sustainable garlic growing methods benefits both your garden and the ecosystem as a whole.

Ongoing Education And Development:

Innovation and development are always welcome in the field of agriculture. Keep up with the latest developments in sustainable farming methods, production styles, and garlic types. Make connections with other garlic lovers, sign up for gardening discussion boards, and go to neighborhood gardening gatherings to share ideas and learn about the most recent advancements in the industry.

Think about recording your experiences growing garlic in a journal. Make a note of what went well and what needs improvement. This contemplation will be an invaluable tool in the developing seasons to come. Try out different fertilizing techniques, play around with companion planting, and modify your strategy according to the particulars of your garden.

Recall that the experience gained along the route is just as important to the adventure of producing garlic as the final destination—a plentiful crop. Your commitment to growing as a garlic grower and acquiring new skills will surely lead to ongoing success and a growing respect for the art and science of garlic farming. Happy expanding!

THE END

Made in United States
Troutdale, OR
06/17/2024

20616954R10037